Saguaro Cacti and Elf Owls

by Louise Crary

Contents

Science Vocabulary

root
A **root** is the part of a plant that takes in nutrients and water from the soil.

A saguaro cactus's **roots** take in water and nutrients quickly.

stem
A **stem** is the part of a plant that carries water and nutrients to the leaves and food back to the roots.

The **stem** of a cactus stores water.

leaf
A **leaf** is the part of a plant that makes food for the plant.

spine

A cactus has a kind of **leaf** called a spine.

flower
A **flower** is the part of a plant that makes seeds.

Saguaro cacti have white **flowers,** which produce fruit.

seed
A **seed** is a part of a plant from which another plant can grow.

Cacti **seeds** form inside cacti fruit.

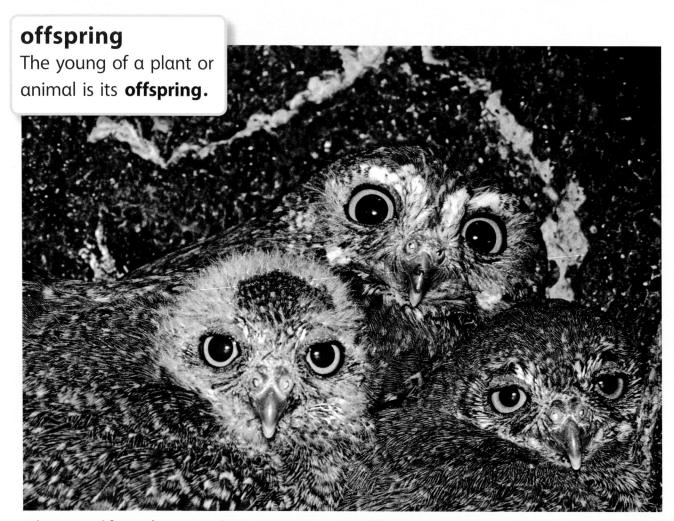

These elf owls are their parents' **offspring.**

variation

A **variation** is a difference in how a plant or animal looks when compared to the same type of plant or animal.

My Science Vocabulary

flower

leaf

offspring

root

seed

stem

variation

There are **variations**, or differences, in elf owls.

Parts of a Saguaro Cactus

Many plants grow in a desert. A saguaro cactus is one kind of desert plant.

A desert is dry. Only certain kinds of plants can live in a desert.

A saguaro cactus's parts help it stay alive in the dry desert. Many plants have the same parts.

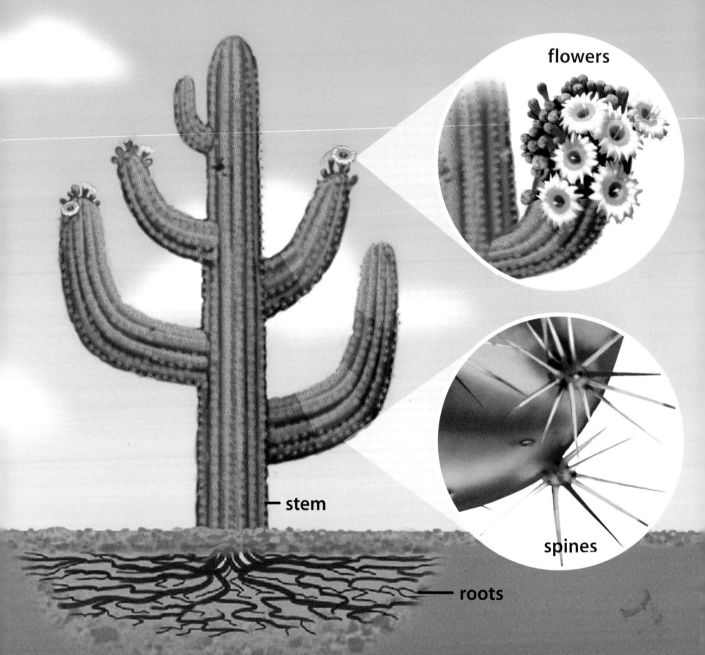

flowers

stem

spines

roots

The **roots** of a saguaro cactus take in rainwater and nutrients from the soil.

roots ——

When this saguaro cactus fell, the roots pulled out of the ground.

root

A **root** is the part of a plant that takes in nutrients and water from the soil.

A **stem** carries water, nutrients, and food to the other parts of a saguaro cactus plant.

A saguaro cactus also uses its stem to make food and store water.

stem

A **stem** is the part of a plant that carries water and nutrients to the leaves and food back to the roots.

Cacti have spines. Spines are a kind of **leaf.**

spine

Spines grow along the stems and arms of saguaro cacti.

leaf

A **leaf** is the part of a plant that makes food for the plant.

White **flowers** grow on the saguaro cactus. A cactus flower can make fruit. The fruit is red. It has **seeds** inside.

flower

A **flower** is the part of a plant that makes seeds.

seed

A **seed** is a part of a plant from which another plant can grow.

Birds and other animals eat the fruit.
They spread the seeds.

This bird uses its beak to
eat cactus fruit and seeds.

A Saguaro Cactus Changes

A saguaro cactus begins as a seed. It grows and changes to look like an adult cactus.

Adult cactus

- In what ways do the adult cacti look like one another?
- In what ways are they different?

Seed

- How are the seeds different from the adult plant?

Seedling

- In what ways do the seedlings look like the adult plant?
- In what ways are they different?

Young cactus

- In what ways does the young plant look like the adult plant?
- In what ways is it different?

Elf Owls

Many different kinds of animals live in deserts. Elf owls live in deserts where saguaro cacti grow.

Elf owls make their nests in saguaro cacti.

Most elf owls look alike. They look like their parents.

This owl spreads its wings. Can you see the many feathers on its wings?

But not all elf owls look the same.

There are **variations.**

variation

A **variation** is a difference in how a plant or animal looks when compared to the same type of plant or animal.

Elf owls have wings with soft feathers. The pattern or color of their feathers might be different.

Elf owls hunt at night. Their excellent night vision helps them catch animals to eat.

A female elf owl can lay eggs. Young owls hatch from the eggs. The young owls are their parents' **offspring.**

adult

This is a mother owl with two of her offspring.

offspring

The young of a plant or animal is its **offspring.**

Young owls look like their parents.
But they are smaller.

Three young elf owls are about
the size of one adult hand.

An Elf Owl Changes

After an owl hatches, it stays in its nest until it learns to fly. In the nest the baby owl grows and changes.

Adult owl

- In what ways do adult elf owls look alike?
- What variations exist?

Egg

- How is the egg different from the adult elf owl?

Hatchling

- In what ways does the hatchling look like an adult elf owl?

- In what ways is it different?

Young owl

- In what ways does the young owl look like an adult elf owl?

- In what ways is it different?

Elf owls and saguaro cacti grow and change in deserts.

Saguaro cacti and elf owls grow and change to look like their parents.

Conclusion

There are many plants and animals in a desert. The saguaro cactus and elf owl are two examples. These living things can grow and change in a desert.

Think About the Big Ideas

1. How are saguaro cacti alike and different from other plants?
2. How are elf owls alike and different from other animals?
3. How do saguaro cacti and elf owls grow and change to look like their parents?

Share and Compare

 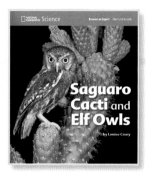 

Turn and Talk

Compare the plants and animals in your books. How are they alike? How are they different?

Read

Find your favorite photo and read the caption to a classmate.

Write

Describe how the animal in your book grows and changes to look like its parents. Share what you wrote with a classmate.

Draw

Draw the plant from your book and label its parts. Share your drawing with a classmate.

Meet Greg Marshall

Scientists use tools and technology. These tools can help them study animals.

Greg Marshall needed a special camera to study animals. So he invented Crittercam. Crittercam records animals' behaviors in the wild.

Index

Acknowledgments
Grateful acknowledgment is given to the authors, artists, photographers, museums, publishers, and agents for permission to reprint copyrighted material. Every effort has been made to secure the appropriate permission. If any omissions have been made or if corrections are required, please contact the Publisher.

Photographic Credits
Cover (bg) Ron Austing; Frank Lane Picture Agency/Corbis; Cvr Flap (t), 4 (t), 11 Matt Meadows/Peter Arnold, Inc.; Cvr Flap (c), 19 Craig K. Lorenz/Photo Researchers, Inc.; Cvr Flap (b), 25 (t) Rick & Nora Bowers/Alamy Images; Title (bg) Jeff Badger/Shutterstock; 2-3, 5 (bl), 14 (bg), 24 (l), 25 (c), 26-27 (bg) Paul & Joyce Berquist; 4 (b), 12 Bruce Dale/National Geographic Image Collection; 5 (t), 13 Bates Littlehales/National Geographic Image Collection; 5 (br), 17 (t) Arizona Sonora Desert Museum copyright 2008; 6, 22 Rick & Nora Bowers/Alamy Images; 7, 20 McDonald Wildlife Photography/Animals Animals; 8-9 James P. Blair/National Geographic Image Collection; 14 (inset) National Park Service, Saguaro National Park; 15 Maslowski Wildlife Productions; 16 (l) Graham Prentice/Shutterstock, (r) Corbis; 17 (c, b) Wardene Weisser/Photoshot; 18, 28 Art Wolfe, Inc.; 21 Craig K. Lorenz/Photo Researchers, Inc.; 23 Lewis W. Walker/National Geographic Image Collection; 24 (r) Anthony Mercieca/Photo Researchers, Inc.; 25 (b) Lightworks Media/Alamy Images; 27 (inset) Rick & Nora Bowers/Alamy Images; 30-31 (insets) National Geographic Remote Imaging; 31 Mark Thiessen/National Geographic Image Collection; Inside Back Cover (bg) George F. Mobley/National Geographic Image Collection.

Illustrator Credits
10 Miro Design

Neither the Publisher nor the authors shall be liable for any damage that may be caused or sustained or result from conducting any of the activities in this publication without specifically following instructions, undertaking the activities without proper supervision, or failing to comply with the cautions contained herein.

Program Authors
Randy Bell, Ph.D., Associate Professor of Science Education, University of Virginia, Charlottesville, Virginia; Malcolm B. Butler, Ph.D., Associate Professor of Science Education, University of South Florida, St. Petersburg, Florida; Kathy Cabe Trundle, Ph.D., Associate Professor of Early Childhood Science Education, The Ohio State University, Columbus, Ohio; Nell K. Duke, Ed.D., Co-Director of the Literacy Achievement Research Center and Professor of Teacher Education and Educational Psychology, Michigan State University, East Lansing, Michigan; Judith Sweeney Lederman, Ph.D., Director of Teacher Education and Associate Professor of Science Education, Department of Mathematics and Science Education, Illinois Institute of Technology, Chicago, Illinois; David W. Moore, Ph.D., Professor of Education, College of Teacher Education and Leadership, Arizona State University, Tempe, Arizona

The National Geographic Society
John M. Fahey, Jr., President & Chief Executive Officer
Gilbert M. Grosvenor, Chairman of the Board

National Geographic School Publishing
Hampton-Brown
www.NGSP.com

Printed in the USA.
RR Donnelley, Atlanta, GA

ISBN: 978-0-7362-7586-6

11 12 13 14 15 16 17

10 9 8 7 6 5 4 3